Viktoriq Vasil'ewna Savelyeva

Transformation of infrastructure systems Part 3

Viktoriq Vasil'ewna Savelyeva

Transformation of infrastructure systems Part 3

Integrative transformation of smart technology infrastructure systems and their constituent parts and elements

ScienciaScripts

Imprint

Any brand names and product names mentioned in this book are subject to trademark, brand or patent protection and are trademarks or registered trademarks of their respective holders. The use of brand names, product names, common names, trade names, product descriptions etc. even without a particular marking in this work is in no way to be construed to mean that such names may be regarded as unrestricted in respect of trademark and brand protection legislation and could thus be used by anyone.

Cover image: www.ingimage.com

This book is a translation from the original published under ISBN 978-620-7-44754-1.

Publisher:
Sciencia Scripts
is a trademark of
Dodo Books Indian Ocean Ltd. and OmniScriptum S.R.L publishing group

120 High Road, East Finchley, London, N2 9ED, United Kingdom
Str. Armeneasca 28/1, office 1, Chisinau MD-2012, Republic of Moldova, Europe
Printed at: see last page
ISBN: 978-620-7-75677-3

Contents

Introduction

The state and technical level of modern special technological equipment and technological equipment and tools allow for the widespread application and integration of artificial intelligence and its functional elements into its control and monitoring systems for use in supersystems and subsystems of production and residential infrastructure.

This circumstance fundamentally changes the situation in engineering and technology, as well as in the standards of design and technological development, applying new complex solutions with already widely and deeply developed professional methods, providing for the use of artificial intelligence elements and artificial neural networks, which in the production process integrate the possibility to work in parallel with the process itself and systematic professional retraining, eventually significantly increasing the production capacity.

The processes of training the personnel of production and scientific units for the introduction of new technological processes makes it necessary to reevaluate artificial neural networks and their functional elements for use in supersystems and subsystems of production and residential infrastructure in the processes of training and professional reorientation.

The complex application of such terms as quantum superiority and its peculiarities for application in functional elements and for use in supersystems and subsystems of smart industrial and residential infrastructure, which emerged in the processes of retraining and professional reorientation, makes it possible to obtain full controllability of processes in combination with a remote form of work organisation.

To increase the level of efficiency of all mentioned processes the most important element is quantum neural network and its features for application in functional elements and for use in supersystems and subsystems of smart industrial and residential infrastructure with high level of automation of all functional processes.

For correct identification of complex system of preparation for organisation of all mentioned processes as practice shows advanced quantum neural network and its features for application in functional elements and for use in supersystems and subsystems of smart industrial and residential infrastructure are necessary.

CHAPTER 1

Technological aspects of building a multilayer three-dimensional light emitter using the method of successive layer-by-layer polymerisation

Takt of technological line operation. Duration of technological transitions:

- transport of the 3D optical light emitter blank from the previous working position to the next working position, transition time 3 seconds;
- setting, pressing the position table for 1 second, pause for 2 seconds, during which all elements of the position are brought into working position;

application of the liquid agent (fluorescent material compounds) - total time is 3 seconds, of which the setting movements of the multi-flare nozzle is 2 seconds, application takes 1 second, removal of the multi-flare nozzle from the working space takes 2 seconds; hot gas blowing takes 1 second;

- the drying and curing process requires a total time of 6 seconds; of this, 2 seconds are required to bring in the radiation shield, 2 seconds are required for the heat treatment itself, and 2 seconds are required to remove the radiation shield;
- marking symbols application - total time is 6 seconds, of which 2 seconds are required for mask feeding, orientation and vacuum pressing, 2 seconds for exposure, 2 seconds for removing the mask from the working volume of the working position;

From the above it is clear that the working cycle (clock) of the technological line should be equal to 3 seconds, for transitions with operation duration of 6 seconds, two parallel working positions should be provided in the line.

CHAPTER 2

To the question of the thickness of the layer(s) of optical material grown in one complete technological cycle

During one technological cycle it is necessary to apply three optical layers, one of these layers, which is located between the other two layers - should be made of light-conducting fibre (material) and its thickness should be within 0,125 mm, limiting it from two sides layers of optically transparent fluorescent material should have each thickness of 0,5 mm, the total thickness of the mentioned construction of three layers is 1,125 mm.

The first advantage of the proposed technology is that the specified thicknesses can be changed, if necessary, without any changes in the design and layout of the processing equipment and using the same tools and fixtures. The second advantage is that, even within a single three-dimensional optical light emitter, the thicknesses of layers or groups of layers can be varied to meet various additional conditions and requirements; a system of combinations of layer thicknesses can, for example, allow the introduction of a specific volumetric geometric code to maximise fluorescence.

CHAPTER 3

To the question of accuracy and geometrical proportions between elements and
surfaces of three-dimensional optical light emitters
manufactured by the proposed technology

It is a system of interrelated dimensional parameters and their limit deviations, their mutual influence and the degree of influence on other dimensional parameters of three-dimensional optical light emitters, when manufacturing a three-dimensional multilayer optical light emitter according to the proposed technology, the average accuracy of all its elements depends on the following conditions:

а)   accuracy of setting the working position on the table;

б)   accuracy of orientation of the optical light emitter workpiece disc relative to the table axis of the working position;

в)   accuracy ratios of manufacturing and assembly of the table and other elements of the working position;

г)   accuracy of weight and volume parameters of the dosage of material, which is applied to the surface of the disc - workpiece;

д)   uniformity in the distribution of material on the surface of the disc - workpiece;

ж)   uniformity of dependence on different types of impact on the disc - blanks and its elements, in the manufacturing process, referred to both linear and volumetric parameters (including temperature variants of impact);

з)   accuracy and homogeneity of the chemical composition of the materials used;

и)   material dosage accuracy and ratio accuracy (weight and volume) when alloying additives are dissolved in base materials;

к)   accuracy of dosage and dissolution of catalysts in base materials;

CHAPTER 4

More info

Comparative characterisation between contact mask and projection mask. The contact mask as applied to the technique and technology of successive layer-by-layer growth of the optical body of a three-dimensional light emitter has the following advantages over the projection mask:

а)   Its use does not require the use of complex optical projection systems;

б)   Its use does not require high positioning accuracy from the equipment components;

в)   the cost of manufacturing a contact mask is significantly lower;

г)   the operating costs of the contact mask are significantly lower;

д)   the required manufacturing accuracy of the contact mask is significantly lower;

е)   mechanical strength and wear resistance of the contact mask is significantly higher;

ж) The contact mask corrects the geometry of the disc when it is pressed against the workpiece;

з)   when using a contact mask, there is no need for complex correction of the coordinates of the mask and the workpiece disc during their identification and mutual orientation;

и)   Due to the use of metal coatings, the life of the mask is quite long, which determines a more efficient use of funds spent on its manufacture;

к)   Due to the fact that the contact mask has a polished contact surface with a metallic coating, adhesion to the polymer layer is very low;

л)   The contact mask has a higher working accuracy because the air gap is eliminated when it is vacuum pressed against the workpiece disc.

CHAPTER 5

Description of the process of building a single layer in a multilayer optical light emitter

Description of the process of building a single layer in a multilayer optical light emitter, which in the described case has the form of a disc with standard dimensions - diameter up to 120 millimetres, thickness - 3.2 millimetres.

CHAPTER 6

General definition

The blank disc of a three-dimensional optical light emitter should consist of a base layer with an electrically insulating thermal protector and an optical working body with at least 5 working layers. Due to the fact that the thickness of all the elements of the working layer is very small - both for the light-sensitive elements and for the optical limiting elements limiting the light-sensitive element from above and below, the manufacturing process of the disc - blank is presented as a sequential layer-by-layer polymerisation of a pre-applied layer of monomers of the material from which each of the elements of the working layer is supposed to consist. Such technological method allows to use the whole range of structural optical materials known at present. In addition, this technological method allows to carry out all operations related to the finishing of the layers also sequentially, from working layer to working layer, which greatly simplifies the process of formation of the light emitter, and, in many cases, makes it generally possible, based on the currently known level of technology.

Thus, each of the working layers consists of 3 elements located in the body of the disc - workpiece sequentially in this order: - optical limiting element;

- photosensitive element;

- optical limiting element.

Due to the fact that the monomer of the material of each of the elements is applied to an already polymerised layer of similar material, there is a short-term diffuse penetration of the monomer into the polymer (not more than 50-75 nanometres), which eliminates the need for adhesive operation and the presence of a parasitic adhesive layer in the disc - blank, which deteriorates the quality of the disc - blank.

CHAPTER 7

Description of the process of building the first layer after the base layer, the working
layer.

Characterisation of the base layer, the material for the base layer can be:

Option 1 - polycarbonate;

Option 2 - organic glass;

Option 3 - organic glass copolymers;

Option 4 is transparent polystyrene copolymers.

- The base layer is intended to be injection moulded and must have a homogeneous structure.

- an electrically insulating and at the same time thermally conductive coating, e.g. made of artificial diamonds, with a thickness of 10 micrometres must be applied on the outer side of the base layer.

- thickness of the base coat without coating -0.59 millimetres, after coating - 0.6 millimetres.

After coating, the base layers are placed in a technological container, similar to the type of cassette previously used in photolithography of semiconductor production with a substrate diameter of 125 millimetres. One cassette holds 25 base layers. The cassettes are gripped by a robot and placed on the loading and unloading fixture of the production line as required. The loading and unloading device removes the base layers one by one from the cassette and transfers them to the conveyor and its guides of the production line, which in turn transfer each of the base layers to the first working position.

At the first working position there is a centrifuge with a vacuum table at the bottom and a system for uniform supply of liquids to the surface above the centrifuge. After orientation, the base layer is placed on the vacuum table and fixed. The system for uniform supply of liquid monomers applies the required dose of homogeneous composition to the entire surface of the base layer, and the centrifuge rotates to equalise the applied layer in thickness. After that the transport system removes the base layer from the centrifuge stage and places it on the vacuum stage of the second working position, where the thermal radiation emitter is located above the centrifuge. In parallel, rotation and thermal treatment are carried out, after which the applied layer is converted into a polymer state. The transport system then carries the resulting disc blank to the third working position. Depending on the design variant of the disc blank, this position can be used for: - printing servo or optical symbols using a master disc;

- offset printing of information and servo symbols by means of a special matrix die and using a homogeneous composition of photosensitive or fluorescent materials as an ink;

- application of a continuous layer of photosensitive or fluorescent material, in the form of a homogeneous composition thereof.

Subsequent process transitions are the same process steps that are repeated for each of the working layers.

The production line tact is 3-4 seconds.

The process is fully automated. The process of preparation of homogeneous composition, including mixing of monomers of photosensitive material with matrix monomers and introduction of initiator into the mixture can be carried out in auxiliary modules of automatic production line.

CHAPTER 8

Non-contact method

Let's consider the methods of control using non-contact method. Among all considered non-contact methods of control, only optical one has found wider application on metal-cutting tools. Optical and non-optical non-contact inspection methods are divided into many. The following section deals with the different types of optical inspection devices. Non-contact methods of acquiring information are also distinguished as contact methods. For diagnostic purposes, non-contact methods are most often used because they have low labour intensity and rapidity. The information extracted by non-contact thermal method is carried by optical electromagnetic radiation in the infrared region. The reason for the absence of colour and sound is infrared radiation due to the energy of vibrational movement of molecules or atoms. The temperature of the object has no effect on the intensity of the radiation. Infrared radiation can be obtained by heating an object, hence thermal infrared radiation is more commonly referred to as infrared radiation. The given fragment is in the stage of design.

Thanks to the test glass, it is now only possible to control the surface of a workpiece with the same radius of curvature. Due to the large number of rings and the application of the test glass on a spherical surface, it is difficult to check local errors. It is worth noting that the existence of a non-contact interference method for inspection of flat and spherical surfaces of optical parts is undoubted. Non-contact interference methods of control of aspherical surfaces are of great importance.

Optical surfaces that are inspected by means of test glasses are the most common. However, in order to obtain reliable results, the test glass method requires direct contact between the reference surface and the surface to be inspected. This is in some cases unacceptable. In some cases it is necessary to increase the accuracy of control, that is why along with test glasses in the optical industry are used various interferometers, non-contact method of controlling surfaces. The disadvantages of non-contact methods of measurement include the desire of the controller to remove contamination from the product poorly. Sometimes they interfere rather insignificantly with the observation of the control process, which is of great importance in optical measuring methods, and even influence the controlled dimension. The second flaw is that, under known conditions, the influence of the surface roughness of the controlled item becomes noticeable. This is due to the fact that the setting gauge has the same plane quality as the controlled product. Determining directions of further improvement of control methods are new nature and content of processes, which concern reproduction of geometrical parameters of parts of aircraft units or assemblies. In the first place among non-contact measuring systems are laser and optical systems, instruments or devices. Nowadays, there is a need for those types of laser measuring systems that meet the specific needs of aircraft construction. In roughness inspection, contact and non-contact methods are used. Non-contact optical instruments use a method in which one or more light strips are applied to the surface or image to be measured. They repeat the surface structure and irregularities. The following instruments have modifications of PPS - based on the principle of light section, PS-shadow gap, MII-interference of light. Roughness is measured on a scale from 0.1 to 160 μm. In optical devices more often measure / tah tg, and can also see the shape and character of irregularities.

Methods utilising optical techniques for non-destructive testing were developed before other methods were used. The human brain, armed when necessary with a simple lens or microscope is an incredibly effective combination for non-contact detection of shape defects and surface cracks in objects. To date, a large number of inspection operations are still heavily influenced by the eyeball of the brain (or part of it) Unfortunately, such inspection methods are very subjective and give rise to difficulties associated with the need to train qualified personnel, establish suitable quality standards, etc. Where control is limited to certain specialised distance measurements, these measurements can be made with considerable accuracy, but provide at best only a partial solution to the problem of complete control. In the remote detection of acoustic vibrations of the surface of an object of control, optical, microwave and sound waves in air can be applied using interference or Doppler effects. Non-contact optical observation of vibrations of the controlled solid mass is performed using an interferometer. After a laser has passed through a translucent mirror on two beams-one from a fixed mirror of the other wave-a specular reflection is obtained. Beforehand, the images are passed through a photomultiplier tube. In the receiving method of control the sensitivity of the method is 500 times less than in the immersion method.

Among other things, the interferometer is a rather complex and massive device with a vibration-sensitive system. It is to be expected that industry will have a new method of non-destructive surface inspection based on the use of holograms. The use of holography in differential interferometry is a special case. A hologram does the same thing, recognising objects recorded on it. But it reacts only at insignificant changes of their optical properties. Most often, quantitative indicators characterising these changes are extracted from the structure and density of interference fringes - such as waves from the object itself or waves (also reconstructed with the help of a hologram) created by superimposing these two sources of oscillations on each other. The current direction will have the possibility of non-contact control of complex raw surfaces (their vibrations, deformation, cracks and changes in reflective properties). Factors hindering the creation of this method are mostly only technical in nature. For example, it is necessary to ensure strong retention and durability of optical elements with interference pattern. In turn, measurement (control) methods are divided into contact and non-contact. In the case of contact measurements, the measuring tip must touch the surface of the measured part, and the nature of contact can be either point or linear. In case of non-contact measurements (optical, pneumatic, etc.) - the measured object is determined without contact with the measuring tip. The most promising in this respect are photoelectric devices, which allow measuring parts in a non-contact manner. Provided that all units can be kept to a minimum distance from the working area. With photoelectric inspection devices, the light flux emanating from the source is sensed by a photoresistor. Differential circuits are often used to create new types of parts, and allow the fluxes to be corrected and keep the screen movement in line with the part dimensions. The source of radiation can be optical quantum generators (lasers). For non-destructive testing by echo and shadow method, in recent years for non-contact ultrasonic ultrasonic radiation with the help of a laser began to use an optical interferometer, and signal reception by means of laser contactless radiation. The wavelength was 148 kilometres.

The autocollimation method is a non-contact method of measuring the rotation angles of a flat mirror. Monitoring of natural gas leaks on pipelines and compressor stations, gas composition of the atmosphere can be carried out by devices of various types. Optical equipment is of great importance in this range, allowing to determine the concentration of various gases in the atmospheric air in a non-contact way accurately and reliably. New generation instruments are specialised optical devices based on hetero-optical pairs. Their dimensions and consistency in the spectrum of light-emitting-photoreceiving diodes allow to sharply reduce energy consumption while reducing the overall dimensions of the device. Photoelectric devices are widely used in combination with optical elements, rasters, diffraction gratings and interferometers. Incandescent lamps, laser tubes and other light sources can act as a source of luminescence. Light receptors are photoresistors, photodiodes, phototransistors and other light devices. Photoelectric devices have a number of advantages: high accuracy, measurement limits (discrete form of the output signal), non-contact method of control. At the same time they are difficult to use due to high cost and environmental protection. Increasingly, new and more significant challenges require that facilities be equipped with quality control methods that can be applied to wafers. As solid state electronic devices become more and more integrated, there is a growing need for new, high-resolution, rapid scientific inspection methods that can be used to objectively characterise the suitability of single crystals or wafers for the tasks at hand. New requirements due to the increased demands on the size and number of defects on the surface of monocrystal plates are made every year. Methods of control by optical or electrophysical methods are already exhausted. The transition to new metrology will be required, and it is necessary to use all possibilities of scanning tunnelling and atomic force microscopy. It is also necessary to use modern methods of structural properties control with submicron or nanometre resolution. At the same time, new inspection tools should be well integrated into the idea of flexible and continuous high-performance automatic production lines. There is a great need for rapid control of wafer surface contamination with metallic impurities, with sensitivity at -10 at/cm. At the end of each of the three sections, we have reviewed the different types of modern contact and non-contact controls. Typically, contact methods use coordinate measuring devices. With the help of a computer or other means of programming, the control systems of these kinds of units are equipped with. Control methods that can be carried out by non-contact means fall into two categories: optical and non-optical. Optical methods most often use televisions and various display systems, but there are also other methods, such as laser. Non-optical systems use an electric field to measure the parameters of an object. Alternative methods of measurement are ultrasonic and radiation. Possible classifications of methods and means of control. For the different methods we can say: Optical systems are most often used for non-contact inspection. These systems use microelectronic means and machine processing of sensor signals. When optimising the performance and reducing the cost of microelectronics or computing products, optical systems are becoming more and more advantageous from an economic point of view. There are several types of measuring optical devices that are used to perform inspection operations. We will consider the following 3 types. Measurements of constant tt slowly changing parameters are most often made with the help of simple methods: mechanical or optical. Non-contact methods are used including pneumatic methods. Most often for measurement of rapidly changing parameters, as well as for control of their dimensions, various electrical methods are used. Their advantages: low inertia and low influence on the object of measurement due to small size of sensors, remote registration of results with the help of special means. The final result of the development of holographic interferometry is the creation of the latest means and effective methods of controlling the shape of optical surfaces, adhesive auxiliary connections to optical. Modifications of electroelements, as well as modes of operation of devices. Holographic methods are non-contact and make it possible to obtain a clear picture of measurement results, but have a number of advantages in relation to conventional interference methods of optical quality control. Only 3 of the elements listed. Secondly, in order to realise control over objects by holographic methods it is possible to use simple optical schemes to the quality of elements of which very moderate requirements are made. It can allow to reduce considerably the cost price of devices of this kind. Holographic methods have several new possibilities, allowing to create high quality measuring devices.

To measure the surface roughness of products made of soft materials, inspectors often have to use special instruments - screwdrivers. Both profilometers and profilographs have inaccuracies that can be explained by the nature of the contact method of measurement. The force P are the parameters of the instrument that primarily determine the amount of distortion when probing a surface. If the radius of curvature of the stylus is equal to g or greater, it is said that deviations from the centre of gravity are used. can be considered at a certain time interval as

The measuring force, depending on the dynamic characteristics of the probing system, the speed of probing and the nature of the profile of the controlled surface, can vary greatly - This fact is taken into account in the design of devices, In modern profilometers and profilographs, due to the rational design of sensors, as well as reducing the speed of probing achieve a significant reduction in the proportion of the dynamic component P,) in the total force P. If the radius of curvature of the needle at the majority of the needle, then the measuring force can be significantly reduced. There is no doubt that in the process of such forces more or less deep scratches will leave traces on the surface of the controlled product. However, everything depends on the material and its hardness. As the analysis in Chapter VI shows, scratches on the instruments can have a different effect on the stylus readings. When the size of the depressions is large and the difference in groping force at the bottom or protrusion is large, the measurement errors are small. Also the value of small size may be in the case of a hollow profile with a large step of irregularities. With narrow microroughnesses, due to the different deformation conditions of the material at the crest and in the depression, the height profile is smoothed. It follows that the softer the controlled fabric and the cleaner the material surface of the product, the greater the reduction. General relations between the data obtained by probing the plane with needles with radii of roundings g   = 10 μm with the indicators of optical non-contact devices and by measuring forces of 2 C. The abscissa axis of the graph shows the purity classes established with the help of optical instruments, the ordinate axis shows the classes obtained by touching with needles having the above mentioned . The curve refers to the theoretical surface of a completely solid body with very gentle irregularities curve- to the surface of products with hardness <20 kgs1mm and the angle of opening of depressions 100°. Curves relating to the surfaces of products made of steel, bronze, etc., are controlled by profilometers having considerable efforts of feeling clean substrates. Consequently, the methods of forming geometrical parameters of products at all stages of aircraft creation change during machine design. Control methods change. They have to be changed. The methods of control based on comparison of parts with rigid size carriers should be replaced by methods of non-contact control - optical, laser and so on. Based on this, the system of automatic shape reproduction and dimensional control is a set of methods for the creation of geometric shapes in aircraft both after design and. Production results. The system covers the following elements machine design of structural elements of the aircraft nitrogen linkage and assembly of assembly fixtures CNC-based manufacturing of parts sequential methods of assembly of units, docking of products by means of levelling. The system is based on a wide application of precision drawing automatic machines, plane and spatial displays, graph plotters, non-contact means of control - laser measuring systems and optical devices coordinate benches of increased accuracy with CNC metalworking equipment with programme control and other means of automated manufacturing of products. To quantitatively determine the roughness of the machined surface, special instruments are used. Non-contact and contact-stylus instruments are used to control surface roughness. Contact probing instruments have the principle of probing the surface to be tested with a needle. Recording is done by a recorder, which has an enlargement of the surface micro profile in the form of a profilogram. From the results of the oscillation, the roughness of the plane being inspected can be inferred. Thus the instrument is called a profilograph. The oscillation of the groping needle of the instrument is transmitted in magnified form by a system of levers to the scale of the instrument, according to which the roughness is calculated. This oscillating movement of the instrument is called a profilometer. The construction begins. At the present time, there are many options for the development of optical measuring systems for non-contact inspection. To demonstrate the general framework of possible principles, we have considered only 3 methods. Shaffer's article is a good one, covering a wide range of existing systems. Certainly, when it was written in 1979, it could no longer be called a new advance in technology. What is bad is the fact that the physical principles remain essentially unchanged. There is a desire to explore three areas in the field of non-contact diagnostics, the acoustic emission method and video television systems for internal inspection of equipment.

The pandemic of viruses and the need for clear control of infection levels has shown that remote control and monitoring (see 3D model of such a control module) provides almost instantaneous control of infections.
the possibility of monitoring the situation, without the need for laboratory analyses.

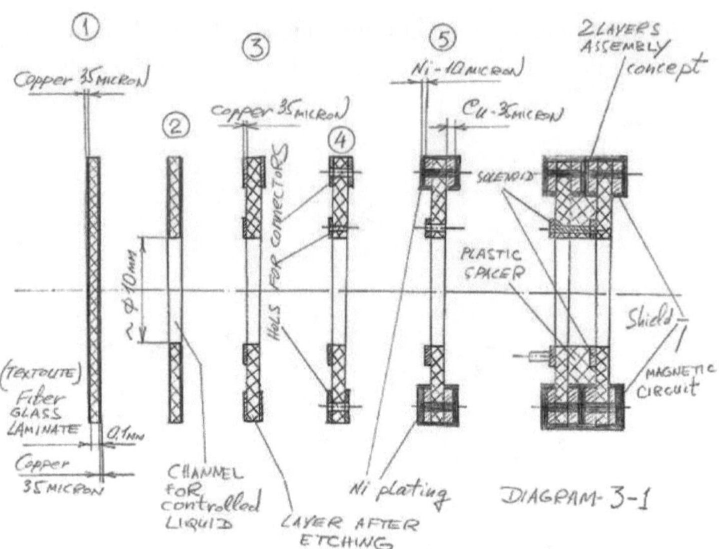

The figure shows the conceptual solution of the layers of a multilayer PCB with electronic noise shielding features.

Non-contact active control of the parameters of the subjects of influence in various modern technological processes (shown in these figures), performed in real time on the basis of the principles of electromagnetic resonance spectroscopy, is also becoming increasingly relevant in innovative production complexes, as well as in various fields of medicine, veterinary medicine, pharmacology, semiconductor and microelectronic manufacturing and food production.

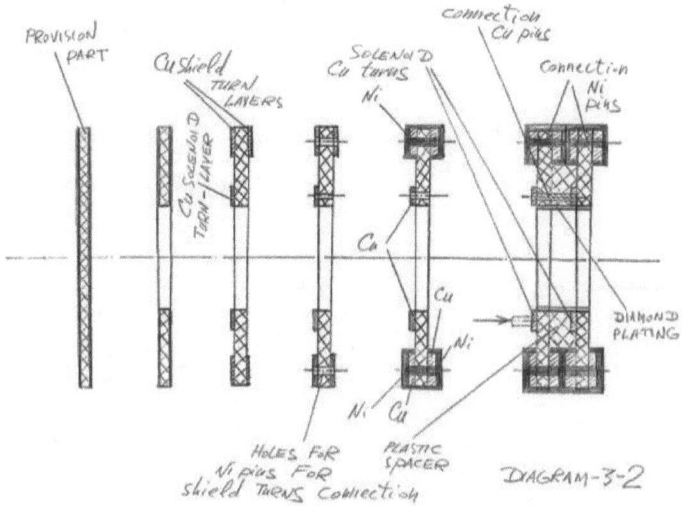

The figure shows the conceptual solution of the layers of a multilayer PCB with electronic noise shielding functions, Figure 2

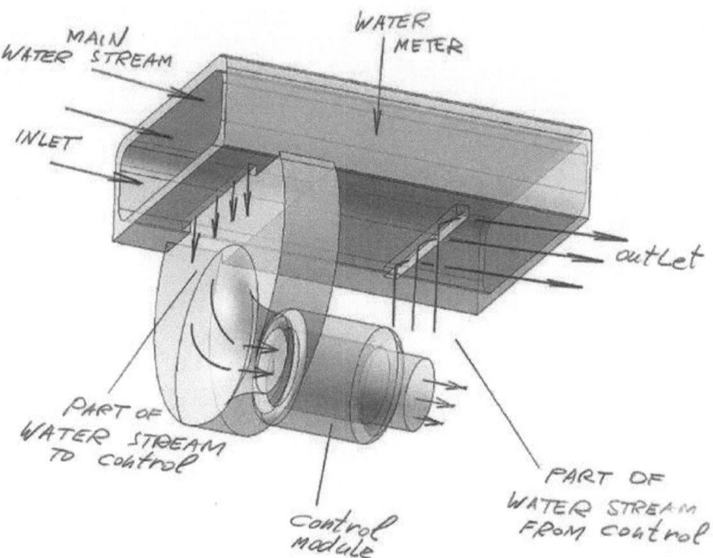

The model shows the industrial principle of connecting a real-time control and online monitoring system in hydraulic lines and pipelines of modern technological equipment.

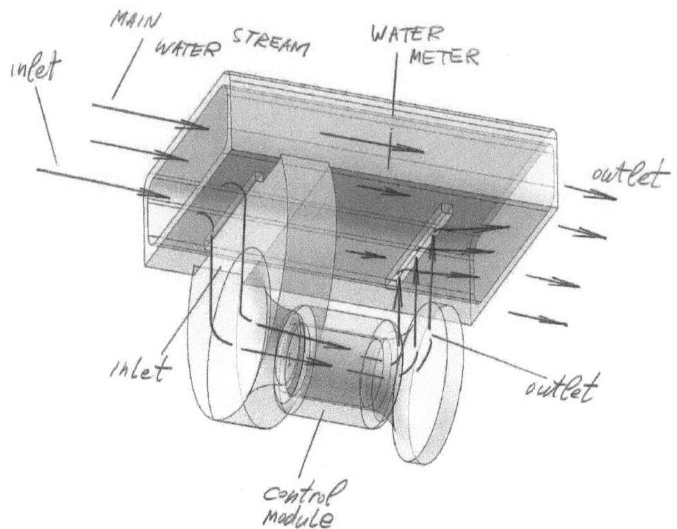

The model also shows the industrial principle of connecting a real-time control and online monitoring system in the hydraulic lines and pipelines of a modern process plant.

equipment figure 2.

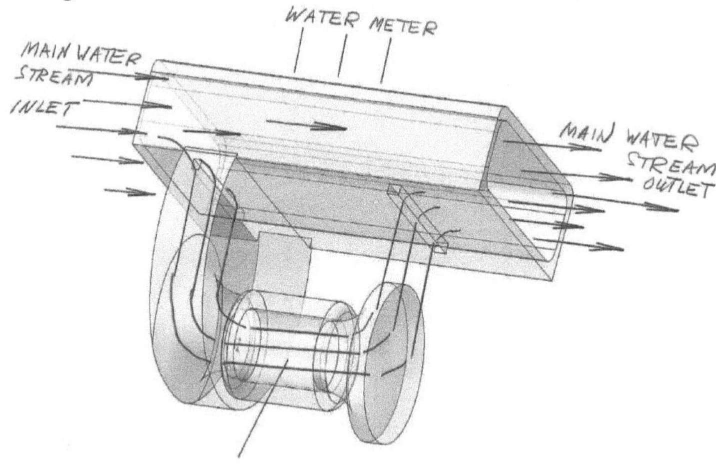

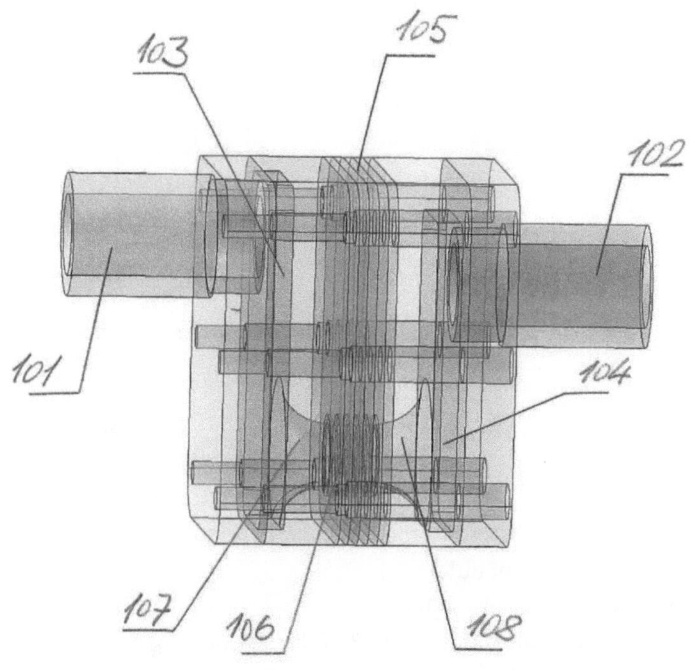

101 - input to the sensor module

102 - output from the sensor module

103 - element forming the input channel of the sensor module

104 - element forming the output channel of the sensor module

105 - sensor in the form of a multilayer printed circuit board

106 - sensor control channel

107 - channel at the inlet to the sensor's communicating vessel system

108 - channel at the outlet of the sensor's communicating vessel system

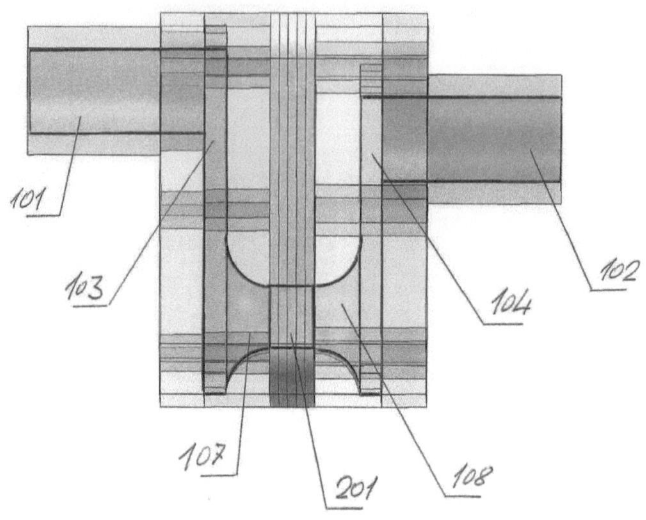

101 - input to the sensor module
102 - output from the sensor module
103 - element forming the input channel of the sensor module
104 - element forming the output channel of the sensor module
107 - channel at the inlet to the sensor's communicating vessel system
108 - channel at the outlet of the sensor's communicating vessel system
201 - sensor in the form of a multilayer printed circuit board

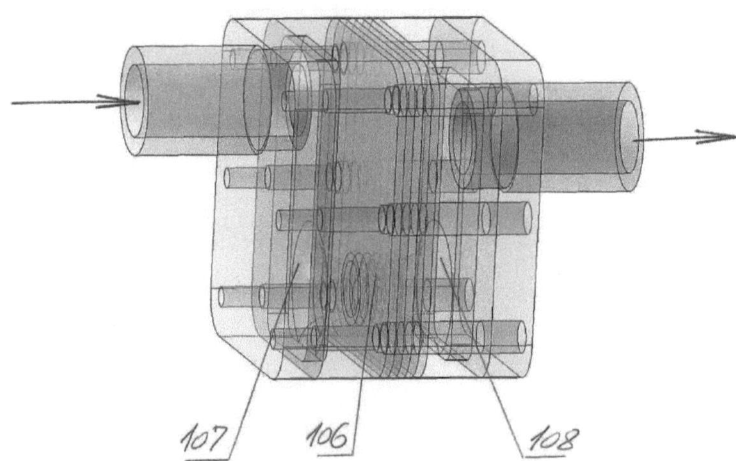

106 - multilayer printed circuit board - sensor with shielding elements and

electronic noise protection

107 - channel at the inlet to the sensor's communicating vessel system
108 - channel at the outlet of the sensor's communicating vessel system

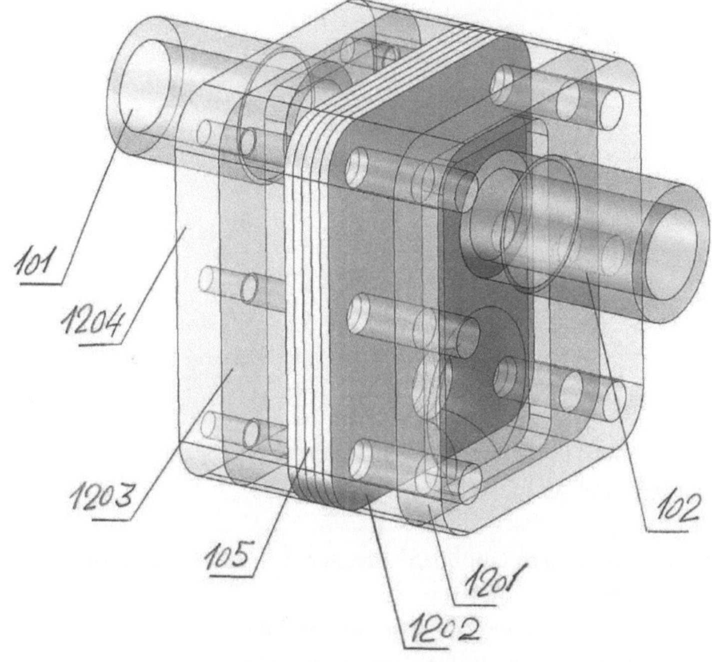

101 - line for introducing the monitored liquid into the sensor module
102 - fluid outlet line after control in sensor module
105 - multilayer printed circuit board - sensor for resonant electromagnetic spectroscopy
1201 - constructional element of the sensor module
1202 - constructional element of the sensor module
1203 - constructional element of the sensor module
1204 - board - constructional element of the sensor module

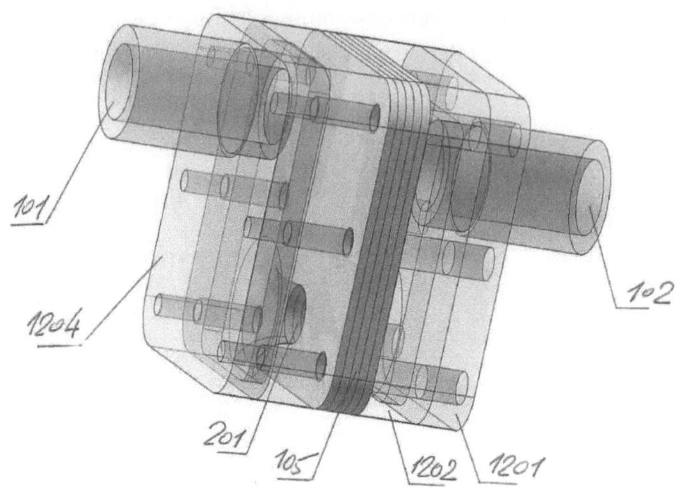

101 - introduction of the monitored fluid into the sensor module
102- output from the sensor module control line
controlled fluid
105 - multilayer printed circuit board - sensor with screen functions
201 - sensor reference channel
1201 - constructional element of the sensor module
1202 - constructional element of the sensor module
1204 - constructional element of the sensor module

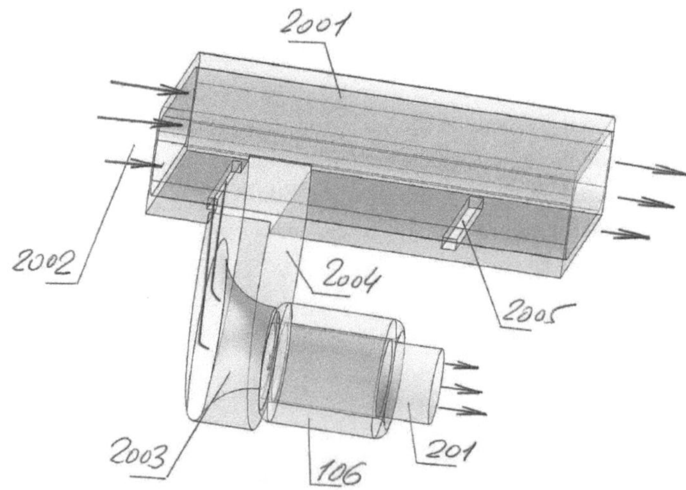

106 - three-dimensional coil (solenoid) of the sensor
201 - fluid outlet line from the control area

24

2001 - **the** main pipeline, in which the flow of the tested liquid moves

2002 - direction of flow of the tested liquid

2003 - inlet channel of the tested liquid flow

2004 - vertical channel creating liquid pressure in the control area

2005 - channel for returning liquid to the flow after control

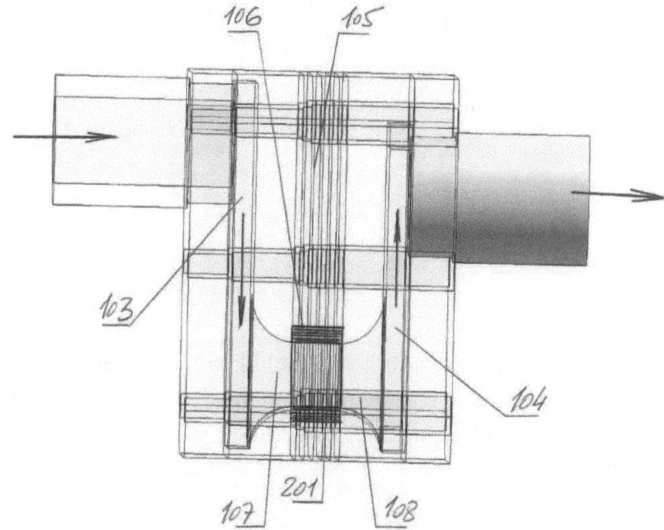

103 - inlet vertical conduit (downward) of the fluid supply system for control in the sensor

104 - vertical outlet (upward) of the fluid outlet system from the control area

105 - multilayer printed circuit board - sensor

106 - solenoid - volumetric sensor coil

107 - supply globoid channel

108 - diverting globoid channel

201 - sensor coil - multilayer printed electronic circuit board

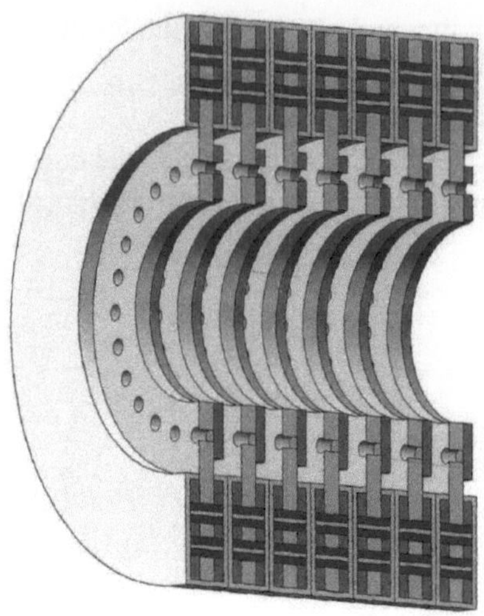

Sensor module in the form of a multilayer printed circuit board with systems of elements acting as a protective screen, reducing the dependence of measurement accuracy on the influence of electronic noise.

**List of used literature, patent and licence
materials**

Annex 1.1.

US Patent   Application20220051358

Species codeA1

Ma; Moses T.   17 February 2022.

METHODS AND SYSTEM OF INTELLECTUAL PROPERTY
MANAGEMENT
USING BLOCKCHAIN

Annotation

A blockchain-based intellectual property management system and methods are presented, which may include one or more *elements* that form an integrated framework for an innovation and intellectual property management ecosystem. *Elements* may include: a distributed intellectual property ledger, a digital intellectual property policy server, non-binary trust models, automatic ontology induction, modifications to the blockchain "mining" and "proof of work" system, an App Store for relevant applications, a partial transparency-accommodating search engine, persistent and encapsulated software trust objects, a smart licensing royalty contract with payment tracking, micro-share incentives, automated The system combines and integrates these functions to provide personal, intra-company, inter-company and extra-company accounting, collaboration, search and its benefits, licensing and tracking of intellectual property information in a networked distributed computing system.

Species codeA1

Sinha; Vineet Binodshanker; and others. 3 March 2022.

BUILDING SYSTEM WITH RECOMMENDATION INTERFACE

Annotation

A construction system including one or more memory devices storing instructions that, when executed by one or more processors, generate a user interface including user interface *elements,* wherein a first user interface element of the user interface *elements of the user interface elements is* associated with a first category and includes one or more first recommendations associated with the first category and one or more first selection fields that allow a user to select recommendations to execute the medium

Species codeA1

Antar, David et al.   **24 February 2022.**

SENSING DEVICE, SYSTEM AND METHOD

Annotation

Variants of the sensor device, method and system utilise a plurality of environmental sensors as a single monitoring and alerting mechanism capable of providing a profile of any pollutant in the form of various gases and particles in the atmosphere expressed in relative concentrations. In various embodiments of the invention, the sensor device may comprise hardware and software *elements including an* electronic control system, a housing, a screen and a cover. Environmental sensors may be attached as part of the electronic control system, and the screen may be shaped to facilitate proper passage of air and sound for efficient operation.

Species codeA1

Brooks, Brian E. ; et al. **17 February 2022.**

DEFINITION OF CAUSAL MODELS FOR ENVIRONMENT MANAGEMENT

Annotation

Methods, systems and apparatus, including computer programs encoded on a computer-readable medium, for determining causal models for controlling an environment. One method includes repeatedly selecting control settings for an environment based on (i) a causal model defining causal relationships between possible settings of controllable *elements* in the environment and responses of the environment that reflect the effectiveness of the control system in controlling the environment, and (ii) current values of a set of internal parameters; and during the repeated selection: monitoring responses of the environment to the selected control settings; determining, based on the responses of the environment, indications of changes in one or more properties of the environment; and in response to the responses of the environment, determining whether the environment is changing.

Species codeA1

Mayuran; Subramaniam; et al. 24 February 2022.

COMPUTATION OF EFFICIENT CROSS-CHANNEL OPERATIONS IN PARALLEL
COMPUTERS USING SYSTOLIC ARRAYS

Annotation

A device for facilitating the computation of efficient cross-channel operations in parallel computing machines using systolic arrays is disclosed. The device includes a plurality of registers and one or more processing *elements* communicatively coupled to the plurality of registers. The one or more processing *elements* include a systolic array circuit for performing cross-channel operations on input data received from a single source register of the plurality of registers, wherein the systolic array circuit is modified to: receive inputs from a single source register at different stages of the systolic array circuit; perform cross-channel operations on channels of the systolic array circuit; bypass disconnected channels of the systolic array circuit, the disconnected channels not being used for computing cross-channel

Species codeA1
Sebastian; Abu; et al.   13 January 2022.
IN-MEMORY PHOTONIC COPROCESSOR FOR CONVOLUTIONAL
OPERATIONS

Annotation

A coprocessor may be provided to perform matrix multiplication of the input matrix with the data matrix in a single step. The coprocessor receives input signals for the input matrix in the form of optical signals. A plurality of photonic memory *elements are* arranged at intersection points of an optical waveguide crossbar array. The plurality of memory *elements are* configured to store data matrix values. Input signals are connected to input lines of the optical waveguide crossbar array. The output lines of the optical waveguide jumper represent a dot product between a respective column of the optical waveguide jumper and the received input signals, and the values of the input data matrix elements to be multiplied with the data matrix correspond to light intensities received at the input lines of the respective photonic memory *elements.* In addition, different wavelengths are used for each column of optical signals of the input matrix.

Species codeA1
XU; Ning; et al.   **3 March 2022.**

ARTIFICIAL *INTELLIGENCE* TECHNIQUES FOR PERFORMING
IMAGE EDITING OPERATIONS BASED ON
NATURAL LANGUAGE
QUERIES

Annotation

The present disclosure includes executing *artificial intelligence* models that determine image editing operations based on natural language queries spoken by a user. Further, the present disclosure includes performing image editing operations using found parameters for image editing operations. Systems and methods may be provided that determine one or more image editing operations from a natural language query associated with a source image, find regions of the source image that are relevant to the one or more image editing operations to create image masks, and perform the one or more image editing operations to create a modified source image.

Species codeA1

LI; Xu; et al.   **24 February 2022.**

SYSTEM AND METHODS OF
ARTIFICIAL *INTELLIGENCE* SERVICE SUPPORT IN THE NETWORK

Annotation

A system including a platform controller for managing an *artificial intelligence service is* provided, wherein the system includes a processor coupled to a memory in which instructions are stored. The instructions, when executed by the processor, configure the platform controller to receive an *artificial intelligence* (AI) service registration request from an AI controller managing the AI service, wherein the AI service registration request includes information about a location of the AI service, and transmit an AI service registration response to the AI controller, wherein the AI service registration response includes routing information at least in part determining how to contact a coordinator associated with the AI service, the coordinator corresponding to the location of the AI service When a request to access the AI service is received from the device, the platform controller is configured to transmit a response to the device, wherein the response indicates whether the request is accepted.

Species codeA1
Sella; Charles Howard; et al. 24 February 2022.
ARTIFICIAL *INTELLIGENCE* SYSTEM FOR THE CONTROL TOWER
AND
ENTERPRISE MANAGEMENT PLATFORM THAT MANAGES THE
LOGISTICS
SYSTEM

Annotation

A value chain system providing logistics system design recommendations typically includes a machine learning system that trains machine learning models that produce logistics design recommendations based on training datasets, each of which respectively identifies one or more characteristics of a relevant logistics system and an output related to the relevant logistics system; an *artificial intelligence system* that receives a request for recommendations; and a digital twin system that receives a request for recommendations based on one or more digital twins of physical assets. A digital twin system that performs modelling based on a digital twin of a logistics environment, one or more digital twins of physical assets.

Species codeA1

Pedersen, Robert D.   3 March 2022.

Artificial *intelligence* expert system *for* motor transport
System and method for preventing
and controlling dangerous driving

Annotation

A specially programmed, integrated system for preventing and controlling dangerous driving of a motor vehicle and methods comprising at least one specialised communication computing machine comprising an *artificial intelligence* electronic expert system capable of making decisions, further comprising one or more electronic sensors of the motor vehicle for monitoring the motor vehicle and monitoring the actions of the driver and/or passengers, including the actions associated with using the

Species codeA1

Brown, John J. et al.   24 February 2022.

FAILURE MITIGATION SYSTEM FOR THE COMMUNICATION
NETWORK

Annotation

In one embodiment, the method includes determining whether a network automation recommendation should be forwarded to a network management system, wherein the automation recommendation determines an action to be taken with respect to a plurality of platforms of the network and is received from a recommendation system that uses *artificial intelligence to* generate the automation recommendation; responding to a decision to forward the automation recommendation, assessing an impact of the automation recommendation; and performing an action with respect to a

Species codeA1
Forzani, Erica, et al.   24 February 2022.

A SYSTEM AND METHOD FOR REDUCING AIR POLLUTION
IN AIR-CONDITIONED ROOMS

Annotation

A system and method for reducing air pollution in a conditioned indoor environment utilises one or more sensor modules configured to detect the presence and/or concentration of particles and/or aerosols at various locations. The control module uses an ***artificial intelligence*** algorithm to selectively activate the at least one pollution mitigation module using programmed machine learning rules and output signals from the sensor module(s). The mitigation module(s) are configured to take one or more actions to reduce the presence and/or concentration of particles and/or aerosols in the conditioned indoor environment.

Species codeA1
Forzani, Erica, et al.   24 February 2022.

A SYSTEM AND METHOD FOR REDUCING AIR POLLUTION IN AIR-CONDITIONED ROOMS

Annotation

A system and method for reducing air pollution in a conditioned indoor environment utilises one or more sensor modules configured to detect the presence and/or concentration of particles and/or aerosols at various locations. The control module uses an ***artificial intelligence*** algorithm to selectively activate the at least one pollution mitigation module using programmed machine learning rules and output signals from the sensor module(s). The mitigation module(s) are configured to take one or more actions to reduce the presence and/or concentration of particles and/or aerosols in the conditioned indoor environment.

Species codeA1

Rai; Vidhi Sandeep17 February   2022.

AUGMENTED REALITY SYSTEM AND METHOD FOR
REAL-TIME MONITORING OF USER ACTIONS
USING EGOCENTRIC VISION

Annotation

An augmented reality-based system and an augmented reality-based method for monitoring a user's actions in real time, wherein the system includes eyewear having egocentric image capture means. A processor and memory are in communication with the egocentric image capturing means. The system captures user activity using the egocentric image capturing means to generate a user activity profile. The user activity profile is then processed using a trained *neural* network to produce a useful activity recognition profile including a set of target actions to be tracked for the user. The system analyses each set of target activities based on predetermined factors to assign each of the set of target activities to one of the predetermined categories. The system provides information to the user based on the *artificial intelligence* analysis of the target actions.

Species codeA1
Blau; ArnaudFebruary 24, 2022

DETERMINATION OF RELATIVE
POSITION OF OBJECTS IN MEDICAL IMAGES BASED ON *ARTIFICIAL*
INTELLIGENCE

Annotation

Methods and systems for classifying first and second objects in an X-ray projection image are described. An appropriate representation and localisation of the two objects is determined by applying models for matching objects in the X-ray image, and a spatial relationship of the classified objects is obtained. Such methods and systems take advantage of *artificial intelligence*.

Species codeA1

Trojoski;   Matthias17 February   2022.

SYSTEM AND METHOD FOR MONITORING THE SCREENING
MACHINE

Annotation

A system for monitoring a screening machine is provided, comprising a vibration sensor configured to record a vibration response of a sieve fabric of the screening machine; and a signal processing device for digitally processing and evaluating the vibration response. In this regard, the signal processing device includes an adaptive algorithm based on *artificial intelligence* techniques associated with vibration responses of one or more comparative sieve fabrics and adapted to characterise the vibration response recorded by the vibration sensor. In addition, a method for monitoring a screening machine is presented.

Species codeA1

Buettner; Florian; et al. 3 March 2022.

TRANSFORMING A TRAINED ARTIFICIAL *INTELLIGENCE* MODEL
INTO A
VALID
ARTIFICIAL *INTELLIGENCE*
MODEL

Annotation

The following describes a computer-implemented method and system for converting a trained *artificial intelligence model* into a valid artificial *intelligence model* by providing a trained *artificial intelligence model* through a user interface of a web service platform, providing a validation dataset that is based on training data of the trained *artificial intelligence model*, generating patterns by a computing component of the web service platform based on the validation dataset, and converting the trained artificial intelligence model into a valid artificial intelligence model. The transformation of the artificial intelligence model is performed by the computational component of the web service platform. The input data, i.e., the trained artificial intelligence model as *well as the* validation dataset, is provided to the computing component via a user interface of the web service platform. Such user interface can be implemented by any applicable frontend.

Species codeA1
Von Mutius; MartinFebruary   24, 2022
METHOD AND SYSTEM FOR PARAMETERISATION OF THE
WIND TURBINE CONTROLLER
AND/OR WIND
TURBINE
OPERATION
Annotation

A method for parameterising a controller of a first wind turbine, wherein the controller sets a control variable of the wind turbine as a function of an input variable. The *artificial intelligence* determines at least one value of the controller parameter for the at least one icing condition/degree of icing of the wind turbine based on a power curve, a load curve, and/or a downstream flow curve of the wind turbine predicted using a mathematical model of the wind turbine for the at least one icing condition/degree, and/or determines at least one value of the controller parameter for the at least one icing condition/degree of icing of the wind turbine

Type codeA1

Bariamis; DimitriosMarch   3, 2022

TRAINING AND MANAGEMENT METHOD
AN ARTIFICIAL *NEURAL* NETWORK CAPABLE OF MULTITASKING,
AN
ARTIFICIAL *NEURAL* NETWORK CAPABLE OF MULTITASKING, AND
A
DEVICE

Annotation

A method for training an *artificial neural* network (ANN) capable of multitasking. For a first flow of information through the ANN, a first path is provided that connects an input layer to at least one task-specific intermediate layer that is common to a plurality of different tasks of the ANN. The first path connects the at least one task-specific intermediate layer to a corresponding task-specific segment of the ANN. First training data is received through the input layer and the first path to train task-specific parameters common to the tasks. At least one second task-specific path is provided for a second information flow through the ANN that is different from the first information flow. The second path connects the input layer to only a subset of the task-specific segments of the ANN, and the second training data is provided through the second path to train the task-specific parameters.

Type codeA1

Mackintosh, Gordon DavidFebruary   24, 2022

A SYSTEM AND METHOD FOR IMPROVING THE
PERFORMANCE OF AN AUTONOMOUS VEHICLE

Annotation

Methods and systems for realising enhanced capabilities of autonomous vehicles. The present invention details an effective and safe methodology for implementing external control and monitoring of autonomous vehicles by authorised personnel, in particular enabling the limiting, controlling and/or disabling of AVs or other mechanisms using non-deterministic *artificial intelligence* (AI) algorithms.

Species codeA1

Lee; Kang Yoon; et al.   17 February 2022.

DC-DC CONVERTER WITH INTELLIGENT CONTROLLER

Annotation

As the inputs of the direct current (DC)-DC converter controller are sampled over a predetermined time and thereby generate two-dimensional state information, wherein one axis is an input physical quantity and the other axis is time, the two-dimensional state information is processed by a convolutional *neural* network to determine and output one of a plurality of control signals. The *artificial intelligence* control portion may operate according to a plurality of operating conditions or dynamically determined operating conditions by applying different *artificial intelligence* engines according to operating modes.

Species codeA1
CHI; Wanchao; et al.   17 February 2022.
ARTIFICIAL ***INTELLIGENCE-BASED*** ACTION RECOGNITION
METHOD
AND CORRESPONDING DEVICE

Annotation

An artificial intelligence-based action recognition method includes: determining, according to video data including an interactive object, node sequence information corresponding to video frames in the video data, wherein the node sequence information of each video frame includes position information of nodes in the node sequence, wherein the nodes in the node sequence are nodes of the interactive object that are moved to perform a corresponding interactive action; determining action categories, with the

Species codeA1
Borrego; Diego A.; et al.   **10 February 2022.**
METHOD AND SYSTEM FOR INSTALLING WIRELESS
SOIL SENSING DEVICES
, MONITORING AND UTILISING THE SIGNALS
TRANSMITTED FROM THEM

Annotation

Several subsurface probes on a plot of land detect soil conditions and wirelessly transmit soil condition information to the service provider's internal server or to a technician's hand-held wireless device. The server provides multiple user interface screens of the client application that display soil condition and irrigation system status information. Local micro weather stations can detect and wirelessly transmit weather status information to the server. The installer may use the client application to determine that the received signal strength on a given probe is weak and should be relocated, or to determine refined location information corresponding to the probe. Historical soil condition information and *artificial intelligence* can predict soil conditions where a previously installed probe has been moved or has stopped working. Probe housings house internal electronics including soil condition sensors, wireless communication modules, processors, and memory. Location coordinates can be acquired at ultra-high speeds to improve location accuracy.

Species codeA1

Borrego; Diego A.; et al.   **10 February 2022.**

METHOD AND SYSTEM FOR INSTALLING WIRELESS
SOIL SENSING DEVICES
, MONITORING AND UTILISING THE SIGNALS
TRANSMITTED FROM THEM

Annotation

Several subsurface probes on a plot of land detect soil conditions and wirelessly transmit soil condition information to the service provider's internal server or to a technician's hand-held wireless device. The server provides multiple user interface screens of the client application that display soil condition and irrigation system status information. Local micro weather stations can detect and wirelessly transmit weather status information to the server. The installer may use the client application to determine that the received signal strength on a given probe is weak and should be relocated, or to determine refined location information corresponding to the probe. Historical soil condition information and *artificial intelligence* can predict soil conditions where a previously installed probe has been moved or has stopped working. Probe housings house internal electronics including soil condition sensors, wireless communication modules, processors, and memory. Location coordinates can be acquired at ultra-high speeds to improve location accuracy.

Species codeA1

Miller; Thomas Gary; et al.   3 March 2022.

ENDPOINT PREPARATION USING ARTIFICIAL *INTELLIGENCE*

Annotation

Methods and systems for realising an endpoint preparation using *artificial intelligence are* disclosed. An example method at least includes receiving an image of a surface of a sample, the sample including a plurality of features, analysing the image to determine whether an endpoint has been reached, the endpoint being based on a feature of interest from the plurality of features observed in the image, and based on the endpoint not having been reached, removing a layer of material from the surface of the sample.

Species codeA1
Aaltonen, Janne ; et al. 3 March 2022.
ARTIFICIAL *INTELLIGENCE* DATA PROCESSING SYSTEM AND METHOD

Annotation

Various systems are proposed to perform tasks related to IPR procurement. The systems utilise a computational architecture capable of providing *artificial intelligence* characteristics. The computing architecture uses a configuration of state variable pseudo-analogue machines, which is implemented by arranging the state variable pseudo-analogue machines in a hierarchical order, wherein the state variable pseudo-analogue machines higher in the hierarchical order are capable of mimicking the behaviour of a human claustrum to perform higher cognitive functions in processing information associated with one or more service requests and to perform quality assurance checks on one or more work pro- cesses In addition, the computational architecture can be realised using a novel configuration of processing devices.

Printed by Books on Demand GmbH, Norderstedt / Germany